Contents

Illustrations

Abstract

This paper analyzes what role, if any, should the United States Air Force take in the nation's stance on planetary defense. Planetary defense in this paper means protecting the planet and therefore our nation from the impact of large Near Earth Objects (NEOs). This topic is usually relegated to the realm of science fiction movies and books but the consequences of such an impact are too great to ignore and simply dismissing it as impossible. Before addressing the role of the United States Air Force in such project, the problem of a NEO impact is analyzed.

First, the paper analyzes the problem itself and discusses the probability that a NEO large enough to cause damage will impact the earth. There is discussion of past NEO impacts during the earth's history and the craters that leave us tell tale signs of these impacts that we can gather data from. Secondly, the effects and consequences of a NEO impact are discussed. This shows the reader how even though the probability may be low, the consequences are so high that we mush take a possible NEO impact as a serious matter.

The paper then discusses how to predict and prevent an impact. In this chapter there is discussion on what the U.S. and other governments are currently doing to prepare for such an impact. We look at how many objects are catalouged that could possibly strike the earth in the near future and what we're currently doing to give the earth advance warning. Also discussed are possbile options to prevent an impact once a NEO is discovered to be on a collision course with the earth.

Finally the problem solution method is used to analyze what U.S. government agencies should have a role in planetary defense, including those that are already contributing or those that do not have a role. The conclusion reached is that the United States Air Force should play an increasing role in planetary defense. The Air Force possesses unique tracking, weapon and space launch capabilities that could greatly aid in the development of a formalized planetary defense system fielded by the U.S. governement.

CHAPTER ONE

On my first cross country move from Vandenberg Air Force Base, CA to Hurlburt Field, FL, my drive across the country took me through the desert of Arizona on interstate forty. I'd planned to do some sightseeing along the way and found that my travels would take me right past the Barringer Meteor Crater near Winslow, Arizona. After seeing countless television shows and blockbuster movies depicting the effects of a meteorite impacting the earth, I couldn't wait to see what a real crater looked like and learn of the actual effects. When I arrived at the crater's rim and took in the panoramic view I was amazed at the size of this scar on the earth's surface.

The Barringer Crater is just over one kilometer across and over 150 meters deep in the center, the collision of an iron-nickel meteorite 50 meters in diameter travelling at approximately fifteen kilometers per second caused a blast equivalent to two and a half megatons of TNT and ejected 100 million tons of rock into the surrounding countryside up to ten miles away.[1] Since this event occurred nearly 50,000 years ago we weren't on hand to witness the devastation at hand.[2] Imagine the physical, psychological, financial, and political effects such a collision would lead to if this happened in our society somewhere on the earth today. It's the equivalent of dropping a two and a half megaton nuclear weapon at random on the earth's surface. In comparison the atomic bomb nicknamed "Little Boy" that was dropped on Japan in WWII had an effective yield of 13 kilotons and warheads in use by the U.S. government today typically have a yield in the range of 300 to 500 kilotons.

As I toured the facilities and spent some time in the gift shop, I picked up two pairs of meteorite earrings for my sisters as a typical "wacky" gift from their brother. It dawned on me that I had the luxury of shopping at this tourist site only because the event had happened fifty millennia in the past. When a similar event occurs in the earth's future, humans won't have this luxury; we or our ancestors will have to deal with the reality of this destructive force. The only way to this avoid this problem is if we start analyzing the problem and doing something about it now. We need to develop a credible system of planetary defense to avoid a catastrophic event in the future and the United States Air Force (USAF) should play a vital role in planning and executing this defense for the United States and eventually for the world as a whole. This paper will demonstrate that despite current efforts towards increasing our abilities in planetary defense, we are still inadequately prepared should the need arise to track and deflect an object that is discovered to be on collision course with earth.

So what exactly are the chances that another impact similar or greater than this will occur in our lifetimes? First, the Barringer meteor crater isn't the only example of an object impacting the earth. The earth's surface is pockmarked with craters created by meteorites much larger and more devastating than the one that created the crater in Arizona. Some are now hidden by oceans or lakes or have been simply buried by the passage of time and some are so large that they can only be noticed from space. Some were not so distant in our past and were close calls like the Tunguska event over Siberia in 1908.[3] This close call involved the air blast of a meteor exploding miles above the earth's surface and the resulting blast leveled an area the size of the state of Rhode Island or a square almost 45 miles long on each side.[4]

Researchers currently estimate there are 170 known impact craters on the earth.[5] Some of these are as large as the 170 kilometer diameter Chixclub crater in the Yucatan peninsula that was caused by a meteorite thought to have been 10-20 kilometers in diameter. We know that the earth is struck frequently, in geologic terms, by meteors and other near earth objects (NEOs) due to the number of craters, but the true numbers may be even larger. NASA estimates that the earth is struck millions of times per day by NEOs that burn up in the atmosphere that are no larger than pebbles.[6] The impact of most of these objects can be neglected because of their small size. However, there are plenty of other objects in our solar system that are large enough to cause major damage if they impact the earth. What is the probability that a meteor large enough to do some harm will survive entry through the earth's atmosphere and produce an impact that causes wide spread damage and destruction? Furthermore, what is the possibility of a meteor impacting the earth that is large enough to cause the extinction of the human race?

The relationship between the size of an object and its chance of striking the earth is inversely proportional. The earth is hit by small objects and debris constantly but a larger object that produces global effects may only impact the earth once every half million years. An impact such as the one that flattened the Siberian forest in Tunguska is thought to happen once every 300 years and it is theorized that it would only occur near a human population center once every 3000 years. Furthermore, these numbers can be extrapolated to tell us that only once every 100,000 years would this impact occur in an urban area and only once every 1 million years would it occur in an urban area within the borders of the United States.[7] These odds don't seem too bad at first, especially considering the last Tunguska sized event was only 100 years ago so we've got some time on the average before the next one occurs.

It's hard for humans to imagine events that occur only once every thousands of years. To compare the time periods between the risks, we have to realize that all of recorded human history only takes us back about 6,000 plus years.[8] However, another factor needs to be added into the equation. With these events we're not merely talking about building your house in a 500 year flood plain and hoping during the next 30-40 years that you live in the house no floods raise above the river's banks. We're talking about devastation on a global scale, possibly one quarter

10

of the earth's population wiped out by a single finite event that lasts only a fraction of a second when it strikes the earth. This type of risk is hard to quantify. Essentially when you look at a larger NEO impact event you are weighing the low probability that a large object will strike the earth with the high consequences once such an object does impact the surface or produces an airburst. In short the risk is much greater when the consequences are much greater and the event in question could wipe out the planet versus destroying a few hundred homes like the major floods we've seen recently in the Midwest.[9]

Some numbers do exist to help us quantify the consequences when tied in with the probability of the risk. Again, we know that the Earth will be impacted by a meteorite at some point, but the key is when? The human mind has a hard time grasping the chances of something occurring during our short life span when we talk in probabilities of an event that occurs only once every hundreds of thousands or millions of years. According to scientists, your chance of dying during a civilization ending impact is greater than the following chances:[10]

- About 300 times greater than the risk of dying from botulism

- About 100 times greater than the chance that you will die in a fireworks accident

- About 10 times greater than the chance of dying in a tornado

- About 1/3 the risk that you will die in a firearms accident

- About 1/30th the chance that you'll be murdered

- About 1/60th the chance that you'll die in an auto accident

Putting the risks in these terms certainly makes it more understandable and implies that the risk of death from the impact of a NEO is real and does exist. In fact, the risk and probability seems even more likely and personal when looking at a list of known NEOs that will pass frighteningly close to the earth in the next twenty to thirty years.

In a little less than twenty years, our usually quiet Earth-Moon system is going to have a lot of visitors. In August 2027, Asteroid Number (AN) 10 is going to get about one lunar distance from Earth. Estimates for its size range from one half to two kilometers in diameter, or plenty large enough to create a regional or global catastrophe if it strikes the earth. Just six months after AN10 passes by object WN5 will get even closer, just about splitting the difference between Earth and the Moon. At 700 meters in diameter this asteroid has the potential for major damage also. By far one of the most famous among the scientific community of end-bringing objects we know about in our solar system is asteroid Apophasis. Astronomers initially thought for a while that this 270 meter-wide rock had an almost 3% chance of hitting us. Since then, odds have been lowered to 1 in 43,000 that it could slam into Earth in 2029. But if it passes through a gravitational keyhole, a tiny region in space that could tweak its orbit ever so slightly, usually where two large object's gravitational pulls effectively cancel each other out, an impact could still happen on April 13, 2036.[11]

CHAPTER TWO

How large of an object does it take to produce visible, damaging effects? There are many factors that influence the effects of the object impacting the earth. The size of the object is an obvious one. Objects from tens of meters in diameter to tens of kilometers in diameter exist in the solar system and could come across the orbit of the earth. The density of these objects also varies greatly. Some meteors are made of an almost solid nickel/iron metallic mix, while others are a less solid stony mass. Also, comets can be made of loosely packed together ice, dust and rock matter. The speed at which these objects can impact the earth range from approximately 12 to 72 kilometers per second or 26,000 to 156,000 miles per hour.[12] The area of the earth where it strikes also has a large effect. Is it near a densely populated area of the planet, does it strike the ocean and create a tsunami, at what angle does it impact the earth? All of these are important variables and questions. For simplicity's sake in this paper many of the variables are averaged and we mainly discriminate between the effects of an object based on the object's size.

The current answer to the question "How big does an impacting object have to be to create major effects?" seems to state that an object with a diameter of tens of meters or greater that survives entry into the atmosphere would be capable of major damage and releasing enough energy to approximate two to three times the power of the atomic bombs dropped on Nagasaki and Hiroshima.[13] When you're talking about kinetic energy being released that equates to the first nuclear weapons developed, it may sound like a lot, but it pales in comparison to the next size up.[14] Real catastrophic effects come into play when the object is in the size range of hundreds of meters, especially as the size of the object approaches one kilometer in diameter.[15] These impacts would produce craters multiple kilometers in diameter and cause destruction on at least a regional scale. The blast energies are in the multi-megaton range and start to exceed the destructive capabilities of the largest nuclear weapons ever produced. At the upper bounds of this category, approaching the one kilometer diameter range, some effects could have global significance with regard to weather patterns and atmospheric cooling similar to a "nuclear winter" type effect.[16] This is due to dust and debris from the impact blocking the sun's rays after it is ejected high into the atmosphere.[17] The effects from a one kilometer wide asteroid with sufficient speed and kinetic energy could impact almost 25% of the earth's population with its effects.[18] Finally, we approach the range where effects are almost beyond belief, the grand daddy of them all.

A NEO in the one to five kilometer diameter range would produce effects that have serious global consequences no matter where on the earth it strikes. Again, a "nuclear winter" type scenario would almost certainly occur where the sun could be blotted out by the dust in the atmosphere for months. Plants, crops and animals would die, the very survival of civilization and the human race would be at stake.[19] Massive earthquakes would work their way across the land. Tsunamis caused by such an event could be hundred of meters high and wipe out entire coastal areas of continents. Acid rain would be created from all the burning pollutants in the blast and freshwater lakes and rivers could be acidified. Even after the dust settled, follow-on effects of increased green house effects could raise global temperatures on an average of ten degrees Celsius for years or decades following the event.[20] Such an impact would have lasting effects on the future of humanity and all the plants and animals on the earth that survived the initial impact.

Another way to measure the potential damage that a NEO can inflict involves the use of the Torino scale. The Torino scale, shown in figures 1, 2, and 3, uses a number from 1 to 10 to represent how potentially threatening an object can be. It takes into consideration how massive the object is, how fast it is moving by listing its potential kinetic energy rated in megatons and on the other axis it rates its probability of impacting the earth from almost zero up to a definite strike or probability of one. This chart makes it easier to discuss and classify NEOs that we can track or identify and their relative danger to the earth.[21] Scientists often use this scale or similar ones to categorize newly discovered NEOs that may be a threat to the earth.

Besides the general scenarios listed previously there are virtually limitless numbers of scenarios that one could construct to illustrate the potential damage from a NEO striking the earth. However, rather than making up another "Hollywood" scenario, it seems easier and more realistic to illustrate the destructive capability of some events that did occur or almost occurred in our earth's past. If the previously mentioned 1908 Tunguska event in Siberia had occurred three hours later, Moscow would have been leveled.[22] In 1908, Moscow's population was approximately 1.36 million persons; imagine the loss of life had this occurred or imagine the loss of life that would occur if the event happened today with Moscow's population of over 10 million. Keep in mind the Tunguska event was the result of a low density stony type of meteor that exploded 40 kilometers in the atmosphere. Had the meteor been a 40 meter wide iron nickel meteorite the blast would have been much worse. A crater 1 kilometer in diameter would have been created and the explosive force would have been around 12 megatons versus the 3 megaton event previously discussed.[23] The threat of impact from NEOs would certainly be treated as a more serious subject had such an event happened in our recent history.[24]

Another pertinent example to be looked at is in the case of something that could occur presently. As this research paper is being written on, an asteroid tagged as 2009 BK58 is passing by the earth within approximately two times the distance from the earth to the moon at a relative velocity of ten kilometers per second. With an estimated diameter of 30 meters and assuming it was composed of some mix of iron and nickel, it would impact the earth and result in a blast of two megatons of energy. It would leave a crater approximately 400 meters wide with a depth of 75 meters and would create a magnitude 5.7 earthquake in the surrounding areas. The explosive amount of two megatons is equivalent to the test of a British nuclear weapon dropped near Christmas Island on 8 November 1957.[25] The blast of a two megaton burst by an object impacting the earth would have the following effects: the heat from such a blast would cause third degree burns fifteen kilometers in radius from the blast, the widespread destruction of structures would occur six kilometers from the center of the blast, and near total fatalities would occur within three and a half kilometers from the blast center. The fireball from such a blast would be at least half a kilometer in diameter.[26] However, as we'll discuss in the following paragraphs, there would be additional effects from this impact.

Thus far the only discussion has been on the physical and direct effects of a NEO striking earth. What about the indirect effects on economies and the stability of nations in a post strike world? Especially the country or countries directly struck or affected by the event? As evidenced by history, nations have succumbed to smaller and less catastrophic disasters. What would stop parts of the world from falling into anarchy and chaos if such an impact occurred?[27] The indirect effects on governments and nations should be taken into consideration when looking at the consequences of such an event. Even if the event occurs far away from our nation, the resulting social and political effects will be felt around the world and will far surpass the range of the physical trauma suffered.

CHAPTER THREE

Now that the actual effects have been discussed, we can see that something probably should be done by the U.S. Government in conjunction with other world governments to identify these NEO risks and prevent them to the best of our ability. In this section we'll address current efforts and if those efforts are enough or if we need to do more. If we have an inbound meteorite the first defense against such an object is time. Time is needed to either destroy it or alter its path, or if none of these options are available, time will be necessary to prepare the earth's governments and population to survive the impact and stack the odds of survival in our favor. Buying this defense of time requires a method to identify and track these objects that might cross the earth's path through space. Currently a system exists and a plan is under way to track and categorize as many near earth objects as possible. These efforts form part of the Spaceguard survey chartered by NASA in the early 1990's. Both optical telescopes and larger radar telescopes are able to track and identify such objects. In fact, most scientists agree that if a more concerted effort was funded and led to identify NEO threats to the earth, we could probably have a 90% chance of gaining twenty years of lead time before an object impacted the earth.[28]

The original Spaceguard survey that started in 1992 had a goal of identifying 90% of the one kilometer or greater objects that could impact the earth and to have identified these by 2008.[29] According to NASA's Jet Propulsion Laboratory website over 5900 NEOs in total have been discovered as of the writing of this paper.[30] Starting with the first object catalogued in 1995, work has progressed steadily and the numbers increase by about 600 new discoveries per year over the past 5 years.[31] Has the goal identifying over 90% of the objects that exist been reached? NASA studies propose that there are over 20,000 NEOs total to discover and identify.[32] So by just looking at the numbers it appears that we haven't reached our goal yet.

However, how do you truly determine when you've reached 90% of an unknown number? It's hard to tell as we'll never know how many objects out there exist; after all, space is a very big place. However, using some common sense math it's doubtful that we've achieved this goal yet. The number of discoveries per year would have to decline and somewhat represent a bell curve for us to think that we were approaching a 90% detection level. (see figures 3 and 4) Detecting around 600 a year means there are too many out there still to slow down our rate of detection.

NASA is working to meet the goal of the Spaceguard survey and is an estimated 75% complete with identifying 90% of the NEO's one kilometer or greater in diameter.[33] The current total of 757 NEO's that are one kilometer and greater in size has added about twenty five new NEOs per year over the past three years. The bell curve on the graph, (see figure 4) is starting to form and this could suggest that they are approaching the point where they have identified the majority of NEOs in this category.[34] In December 2005 the U.S. Congress and U.S. President upped the ante and passed a NASA authorization act and charged NASA with detecting and cataloguing 90% of the 140 meter and larger NEO's by the year 2020.[35] It seems that the US government has realized the importance of this task and realizes while it is nice to know the status of the one kilometer "planet killer" asteroids, it's time to move onto the next step and identify the medium sized "destroyer of nations" asteroids. This second survey could cost more than one billion dollars over the next eleven years if all the projects recommended by the scientists are funded.[36] However, the U.S. government is obviously not taking this threat seriously and has only committed funding to the tune of about four million dollars per year. More funding is needed to continue this task that has been handed to NASA. As of Jun 2008, NASA has ID'd 959 asteroids that will come within 20 times the distance from the earth to the moon. Of those, 5 will come between the earth and the moon in the next century. The U.S.

government and NASA can't rest on their laurels now. The U.S. needs a committed and well funded effort to continue these operations and guarantee advance notice of an impending strike by a NEO.

We can never detect and identify all of the NEO's that could impact the earth. New NEO's are being created outside and inside our solar system constantly. A collision in the asteroid belt or somewhere else in the solar system could create debris and form new objects that are on a collision course with earth. Similarly, the orbit of a current object could be changed by a collision or a piece of it breaking off, or its orbit being adjusted by simply passing near a planet or another object that tugs on it just enough with its gravitational field. Finally, there are comets or other objects that come into our solar system that are out of the view of our optical telescopes until they get too close for us to mount any defensive measures unless we are prepped ahead of time.[37] A simple investment increase of tens of millions of dollars per year or even in the single digits would give the U.S. and the world a much higher awareness of the space surrounding our planet and our potential dangers associated with impacting objects. If the four million dollars spent per year so far as part of the Spaceguard survey has allowed us to increase our knowledge from basically zero to now almost 75% of the objects that might impact our planet and cause catastrophic damage, think of the increase in knowledge we'd have of NEOs from a slight increase in a few million dollars. In the government budget millions of dollars easily drop off the radar scope of most legislators. Most "big ticket" items cost in the hundreds of millions or billions of dollars. Most people who are aware of the problem would agree that we can afford the extra expense, even in today's economy and in these times of shrinking government budgets. In fact, most people aware of the threat would say that we can't afford not to increase efforts with the stakes so high. For example, the entire budget of NASA is approximately $20 billion dollars in fiscal year 2009. For the same period the U.S. Air Force's budget for space as the Department of Defense (DOD) executive agent for space is around $18 billion dollars.[38] Adding $10 million dollars per year to either of these budgets to track NEOs that are capable of causing a regional catastrophe or the extinction of the human race is drop in the bucket compared to the value. Even if Congress directs NASA or the Air Force through DOD to fund this ten million out of their own pocket, it would only account for .05% of their total space budgets. In some cases we probably have spent more government money in study the mating habits of the fruit fly that we have to ensure our own species' survivability on this planet.

These numbers do not include the additional cost for developing a means to "attack" or defend against an incoming object, but once we know an impact will occur without any action on our part, is there really any price or cost that we wouldn't be willing to pay associated with our survival? Even with a regional impact that would occur outside the U.S. borders, the cost of providing disaster relief and the loss of lives and real property that the U.S. relies on for manufacturing and raw material trade with such a nation would make it worth our while from an investment perspective to avoid future costs. This is not even taking into consideration the moral responsibility a nation such as the U.S. must feel if we had the power to stop such an act from occurring. If we spend all this extra money on detection and tracking capability, then what good will it do us if we discover an impact is imminent and are powerless to stop it? What exactly are our plans and options when this problem does present itself, as it eventually will?

First, from a planning perspective, no nation has a detailed or acceptably thought out or tested plan on how to deflect or destroy an incoming object. We are finally getting somewhere in the tracking and identification portion of the problem as mentioned previously, but we are a long way from solving the problem. Most plans deal with mitigating the damage from an impact by preparing our population for the event through the civil authorities and with sheltering in place measures. The typical response would be to sit in old nuclear fallout shelters, distribute food and water and wait out the effects if possible and provide help and aid to those in need who are most affected by the events in other nations once the initial strike has passed. This scenario involves giving up on the most obvious solution, deflecting or altering the trajectory of the body set to impact the Earth. Even though the task of deflecting a large NEO poses complex technical problems it is surely better than just continuing on and waiting for the inevitable. Here the maxim "an ounce of prevention is worth a pound of cure" surely makes sense.

Options for tackling this problem range from obliterating the object with nuclear warheads, slowing it down or deflecting it with pulses from a laser, attaching a solar sail to slow it down, using a gravity tug system to dislodge its orbit, or drilling into its surface to break it apart. Although none of these methods have been proven or even partially developed except in theory, some stand above others as valid courses of action when viewed by engineers, scientists, and space experts in civilian and government circles.

Attacking the most obvious answer first, why can't we just send up a swarm of intercontinental ballistic missiles (ICBMs) with nuclear warheads to intercept it and blow the object apart and completely destroy it? One of the first problems is current ICBM delivery and targeting methods are ground based and current missile configurations wouldn't allow a warhead to travel far enough into space to impact the object far enough away from earth to make a difference in time.[39] The warheads would have to be placed on a modified ICBM body or spacelift vehicle with enough thrust to place the warheads on an intercept path that would be far enough out to make a difference in the objects trajectory. This may not take a great deal of effort when compared to other options but some major modifications in the ICBM guidance and targeting system would be needed. Additionally, the Hollywood movie script of blowing up the object to smaller chunks isn't an ideal plan as that just would create more impactors with roughly the same center of mass that were on a collision course with earth, and now the earth's population has multiple impacts and to deal with. Some studies estimate that destroying an object could create a blast area or area of effects up to two times the original objects damage. For example, a ten megaton blast from a single meteorite would affect an area of 750 square miles with moderate damage. However, by spitting that ten megaton meteor into four two and a half megaton sized meteors with impact areas 20 miles apart, the total area of devastation would equal over 1200 square miles or almost twice the area.[40] This is not on the desirable side of the effects if we are to try and limit the damage or avoid it altogether.

In this situation might not apply however, if the object has a low density and would fracture easily. For example a comet is usually composed of ice and dust and some rock matter loosely held together by its gravitational attraction. If a nuclear blast could break it up into chunks smaller than 20 meters in diameter, this might work. These low density objects smaller than 20 meter in diameter rarely make it through the atmosphere and are usually ablated or explode in the upper reaches of the atmosphere where their effects aren't felt on earth.[41]

In another case using nuclear weapons detonated near the object to affect its orbit rather than trying to completely destroy it could be successful in theory and practice. Scientists have theorized that exploding a nuclear weapon next to an object in space might be the best solution. The radiation and blast would vaporize a small part of the surface area of the asteroid that was exposed directly to the blast. This mass of this vapor would then impart a small nudge or push to the asteroid in the opposite direction of the blast[42] This is only a simple description of the operation required to move the object from its orbit. Complex calculations are required to determine the exact distance a weapon would need to be exploded to create the largest effect. Too close of a blast and you might accidentally fracture the object or the blast would be too concentrated in one area and not spread its full pressure over the surface of the asteroid.[43] If the explosion occurred too far away then the energy imparted would be spread too thinly over too large of an area and wouldn't vaporize enough mass to impart enough momentum on the object to move it.

Another problem with any solution that intercepts the object is accuracy. It's hard enough to get two objects to intercept each other when they're both moving at 17,500 mph in low earth orbit or in the case of intercepting a missile with an anti ballistic missile battery or something similar. We've tackled both of the problems listed above with our anti-ballistic missile system and testing of anti-satellite weapons in the past. However, the problem with intercepting a NEO in bound for the earth is that the object you're trying to intercept is moving at a speed between twelve to seventy two kilometers per second. This is much faster than any incoming ICBM or a satellite that is moving in low earth orbit. Add to this that the object is not coming in on a straight line and your missile or interceptor is not moving towards that target on a straight line either.[44] Both objects are on a slight parabolic arc (when viewed from the macro level) because of the gravitational pull of the sun, earth and any other bodies that are nearby in our solar system. Even though gravity may seem to work against you in some of these situations, in the following deflection operations gravity is the basic force that makes them work.

One option of moving the object out of earth's path involves using a gravity tug to slow or alter the NEOs path. If the object is intercepted far enough away from earth, its orbit may only need to be slowed .1 meter per second. Placing a large satellite or perhaps even a nearby asteroid on the same path as the object and keeping it there will provide a slight but sure tug on the object. The force of gravitational attraction between the two objects will slowly pull the target object off course by fractions of inches.[45] If the object is intercepted in time, these fractions of inches will add up and the object should miss the earth. These are only a few of the options available and the options seem to be only limited by the imagination of the scientists that continue to theorize about them. However, the purpose of this paper isn't to produce an exhaustive list of options, but to provide a background for the reader to sufficiently understand some of the problems and nature of such a large task. Some of these options are listed to provide insight into the progress we've currently made or possible areas we should focus our efforts.

All of the previously discussed items are in theory only. None of them have been tested outside of a laboratory and most theorizing has been done in the scientific community without much funding or official input from the Department of Defense or NASA. Even though the preponderance of knowledge on the subject rests with the civilians, at some point the government will have to get involved due to the nature of the threat and size of such an ambitious project. In the past there have been some tests of rendezvous with NEOs but not for the purpose of testing the options we've just discussed. The European Space Agency (ESA) plans to send some craft to meet up with asteroids for both study and a deflection exercise. The Rosetta project involves an ESA vehicle orbiting, studying and mapping the makeup of an asteroid. The more ambitious ESA "Don Quixote" project involves two vehicles, one to sit on orbit and provide measurements while a second craft impacts the asteroid at high velocity, resulting in measurements to determine the effect on the asteroids orbital parameters is planned.[46] NASA had a joint project with the DOD titled "Clementine I and II" that planned similar tests. One Clementine project was line item vetoed by President Clinton. It was thought this was done because the same technology that could be used to intercept an asteroid or NEO could also be used to intercept an ICBM, and the development of such could violate the Anti Ballistic Missile treaty that the U.S. was held to at the time. The second part of the Clementine went on to map the surface of the moon but soon failed after this and was unable to continue its mission to intercept and study a near earth asteroid.[47] To prove that this capability exists and to further testing efforts international bans on nuclear weapons testing in space would have to be relaxed or ignored.

Testing a complicated system like any of the previously mentioned solutions is invaluable because the amount of effort and lead time dedicated to a destruction/deflection mission of a NEO is so great that we might only have one launch or one opportunity. A reason for the long lead time needed is illustrated in this example. If an amount of one megaton of blast energy is needed to deflect an object when it is ten years away from impacting the earth, just a one year slip in the launch time that you can reach the object to deflect it results in 100 times greater energy needed to deflect the object at nine years out versus deflecting it when it was ten years out from earth. Waiting one year causes you to use 100 megatons of blast energy versus one. Having a system that is redundant is also desirable. If one system fails at launch or fails en route to the rendezvous with the object, there may not be enough time to launch a second one with enough energy to move the object unless a second interceptor is launched or ready on the pad at a moment's notice. Even taking it a step further and launching interceptors with redundant capabilities in sets of threes would seem to limit the chance of complete failure to an acceptable level (if there is such a level when speaking of these types of consequences).

Once your interceptor reaches the target how do you ensure and validate that it is working besides the fact that it has arrived on station? How many years or months does an organization have to observe miniscule changes in orbital parameters to ensure that the object's velocity is being slowed by tenths of meters per second, or how can you ensure that it's heading is being changed ever so slightly and can be measured to the thousandths or millionths of a degree? New technology or new methods must be developed to provide accurate and timely feedback to ensure the system is working. The stakes are too high and one cannot sit around and wait and assume that the orbit of the object is changing and being affected while we waste valuable time not trying a secondary method or launching a backup interceptor.

CHAPTER FOUR

The previous chapters list a few of the problems that need to be worked out for a planetary defense system. Such a system is so complex and will end up being so large and cost prohibitive that it must be a government funded and run organization. In fact, since it benefits the whole world and many countries, it will probably end up being an international consortium of nations who all share in the common fate of our earth or the particular region where the object is predicted to strike. This would allow for a larger funding base and multiple nations of engineers and scientists tackling the largest and most serious problem humankind has seen to date. In 1992 a NEO interception working group meeting at Los Alamos National Laboratory in New Mexico came up with some cost calculations to develop and implement a planetary defense system for three different sizes of NEO's. For small NEOs less than 50 meters in diameter the cost benefit analysis yielded results of one million dollars per year. This is based on the formula for a cost/risk analysis of the damage caused by a NEO impact divided by the number of years between impacts on the average. For a larger NEO around two kilometers in diameter, the cost per year rises to two hundred million dollars and it increased at a geometric rate onward for larger NEOs.[48]

These average cost answers still don't tell us the total cost of a defense system. Estimates show that around 120 million dollars a year over 20 years would maintain a planetary defense system.[49] This does not include the estimated billions to develop the initial system and the development cost depends on which method of destruction or deflection is chosen. Cheaper options include the already existing technologies of nuclear weapons with modified ICBMs launch systems. It would cost substantially more to develop a new technology like a solar sail solution or some type of gravity drive system.[50] It is also estimated that spending this much on a system would only give the earth a limited mitigation system. If we are to develop a system to destroy and deflect those giant five kilometer diameter asteroids that exist those costs could run into the hundreds of billions or even trillions and be the most expensive project in the history of the human race.

As noted by some researchers though, initiating a defense system now before we fully understand the problem would end up wasting more effort, money and time and jeopardize further funding. They theorize that the best approach at this time would be to spend more money studying the best way to track and defend against these NEOs before enacting an actual plan and developing and building new technologies and physical capabilities. With that said, a partial effort to develop some of the new technologies that will be required couldn't hurt the cause at this stage of the problem and solution development. The lead time for developing some technologies could be many years or even decades and starting earlier can only help us in the endeavor.

This brings us to our main question. Who in the U.S. Government needs to shoulder the immense task of managing the project, running the acquisition issues and the funding of such an enormous task? An obvious answer to some would be NASA. They deal exclusively in space and have great experience with launch systems and vehicles and have the necessary engineers and people already on staff to start the task. They also have the current contracts and links with the appropriate civilian agencies and industry that could help the project along via contracted help. Still others would argue that the obvious choice would be the DOD and possibly the USAF. The USAF is already the executive agent for space within the DOD and they already possess the sophisticated satellite tracking capabilities along with the only ICBM force which includes a nuclear capability. The Navy possesses nuclear capability but the Trident launch systems do not have the lift or range capability. The USAF also possess a heavy launch force and have multiple launch vehicles that could be modified or suited to deliver and interceptor an object. Furthermore they have experience with multi billion dollar weapons systems and procurement and acquisition processes. The AF portion of the DOD budget is also already the largest in the nation and could more easily swallow or absorb the funds or excess costs related to such a project.

Another camp would see that the newly formed, post 9/11 Department of Homeland Security would have purview over such a project. After all, the main reason we are concerned about a NEO impact event is to take care of the civilians at home and prevent loss of innocent life. This department would certainly have a valid claim to stake in this case. However, the department of Homeland Security is among the newest of U.S. organizations they may not have the capability and experience within their ranks to develop and maintain such a large and complicated system. They also have virtually no experience in space or weapons systems when compared to the USAF, DOD and NASA.

Whoever gets the final nod to go ahead with the project of planetary defense, chances are there would be a lot of debate in Congress as to who gets the funding for such a project and some of it probably won't be altruistic in nature. Senators and Representatives could be swayed to vote for a certain agency to take the lead depending on the economic impact it would have in their districts. This problem is exacerbated because we're talking about building a system before we need it. Most people still wouldn't see a NEO as a serious threat and the elected representatives may only see this project as a "cash cow" to argue over on Capitol Hill and play politics within their districts. To this day there are still Congressional "food fights" that we see over who gets to build the next tanker aircraft for the USAF or the next search and rescue helicopter contract for the AF. These "food fights" in Congress don't really take into account what the organization knows that it needs or what it wants for the mission. We'd have to be sure such a problem doesn't derail a planetary defense project before it is even started.

Once our chosen organization begins working on the project, who do we want to partner with in the world arena? Certainly many countries would like to help either because they see the true need for a planetary defense system or they see it as a public relations and media bonanza. What developing nation wouldn't want to be seen as contributing to the defense of all human kind and carrying that banner forward in front of their people and regional or neighboring nations? Who would we let in on the project? Allies only who possess the technology and funding to actually help, or anyone who is willing to join with us in this endeavor? Does this include our sworn enemies who might use this opportunity to gain some technological knowledge or advantage over us? What a perfect opportunity for industrial espionage or stealing proprietary information from top U.S. defense contractors. Or could it follow the model of international cooperation that we've seen on the International Space Station? One of the nations that we're currently not on the best of terms with possesses the most powerful launch and booster systems on the planet. Russia's Energia launch system was capable of boosting a payload of over 40,000 pounds into low earth orbit.[51] This is approximately three times the payload that the U.S. Space Shuttle can boost into orbit, and the Space Shuttle is the second most powerful launch vehicle on the planet.[52] Some of our fellow countries also possess or have access to some of the best launch sites on the planet, **Kourou, French Guiana; Alcantara, Brazil or movable** Sea Launch facilities as part of an international consortium.[53] Many of these sites would be advantageous because their close proximity to the equator means an energy savings on launch and would allow a vehicle to carry heavier payloads. Clearly many political considerations outside of the realms of technology and science need to be taken into consideration. The scientists and engineers will have to work within the boundaries created for them in this project by the politicians. However this is nothing unique, science and technology in the modern age has always been held prostrate to the needs and wishes of the state or politicians.

CHAPTER FIVE

With all these problems, possible solutions, and political considerations, where does this leave the future of planetary defense and what is the U.S. Air Force's role in this possible mission? We've established that the threat of a NEO impacting the earth is real based on the probabilities the scientists have computed and the evidence of past strikes on the earth and the associated consequences. We've also established that in order to prepare for an impact from a NEO we need to track and identify these objects to gain valuable warning time. The USAF can further assist with this effort. Although NASA seems to be doing fairly well in the Spaceguard survey, their efforts can probably be helped by the Air Force Space Surveillance Network (AFSSN). The AFSSN maintains a worldwide presence of radar and optical based tracking stations that virtually cover the globe. These assets could be used to augment the NASA work and confirm some of their observations. A lot of the AFSSN capability remains in the classified arena and cannot be discussed here but some efficiencies could be gained here. The AF also controls most of the satellites in orbit or has a liaison capability with the National Reconnaissance Office (NRO) who controls the NRO satellites. Some of these assets could also be used to track NEOs that seemed to pose a larger than average threat to the earth. The USAF would be in a better position than any other agency to liaise with the NRO to accomplish this.

We've established that the USAF could help more than it already is and increase the fidelity of tracking by working alongside the NASA Spaceguard survey. How might the USAF help in actually deflecting and/or destroying a NEO whose impact with the planet was imminent? The USAF is uniquely qualified to lead or assist in this effort as they are the only organization who has both the stewardship of various types of nuclear weapons and also owns and maintains the launch vehicles to put them into orbit via modified Minuteman, Peacekeeper, Titan or Delta

Evolved Expendable launch vehicles. Some of the USAFs lack of experience at intercepting objects in deep orbit would provide a good partnership opportunity with NASA. USAF launch facilities that exist at Vandenberg AFB in California and those that exist at Cape Canaveral in conjunction with NASAs launch sites in Florida could assist and provide the capability to launch against a single or multiple targets in rapid succession. Many people don't realize that many of NASA and other government agencies have their payloads placed in orbit via USAF launch systems and vehicles.

Furthermore, the USAF has a large space acquisition community and regularly works with many of the commercial space contractors in the nation and internationally. Bringing this acquisition and management experience to bear on the problem of managing a multi billion dollar contract such as the planetary defense system for the U.S. will free the engineers and scientists involved in NASA and other organizations to focus on the technological advancements required for the system to be workable and cost effective. Based on these above general criteria it is essential that the USAF be intimately involved with such a project and co-lead it with NASA while other US organizations could work under both lead agencies. NASA and the USAF have shared an interest and have worked in the environment of space since John Glenn first orbited the earth. NASA's mastery of the space environment and the USAF mastery of weapons applications would dovetail perfectly for the project of planetary defense. To cut either of these organizations out of the loop would seriously hamper any progress in this area.

Regardless of how the U.S. Government proceeds in establishing a plan for planetary defense, it's important that the subject is given the appropriate level of attention and the world finally takes steps to protect itself from possible impacts. Much work remains to be done in the areas of tracking, prediction and finally deflecting the objects that pose a serious risk of impact.

Taking the first step by delegating this task to the USAF and NASA as co-executive agents for the task will set us on the right track to solving this problem. Man has proven time and time again that almost no challenge is too great or formidable for our imagination and application of our full effort. This can be proven again if the U.S. and world governments set us up for success by picking the proper leadership structure to tackle this important problem.

[1] Barringer Crater Website.
[2] Meteor Crater Enterprises Website.
[3] Urias, *Planetary Defense,* pg. IX, Executive Summary.
[4] Barringer Crater Website.
[5] Hamilton, Solarviews Website.
[6] Sharpton, World Book Website.
[7] NASA *Spaceguard Survey,* pgs. 32-33.
[8] Lewis, *Rain of Iron and Ice*, pg 185.
[9] Young, *Chances Small for Head on Collision with Killer Asteroid,* pg. 1.
[10] Knox, *Planetary Defense Legacy for a Certain Future*, pg. 17.
[11] Harvard Smithsonian Center for Astrophysics Website
[12] Lewis, *Rain of Iron and Ice,* pg. 37.
[13] Knox, *Planetary Defense Legacy for a Certain Future,* pg. 10.
[14] NASA *Spaceguard Survey,* pgs. 26-27
[15] Knox, *Planetary Defense Legacy for a Certain Future,* pg. 11.
[16] Knox, *Planetary Defense Legacy for a Certain Future,* pg. 11.
[17] NASA *Spaceguard Survey,* pg. 28.
[18] Urias, *Planetary Defense,* pg. 8.
[19] Knox, *Planetary Defense Legacy for a Certain Future*, pg. 17.
[20] NASA *Spaceguard Survey*, pg 29-32
[21] Kowitz, *The Department of Defense Take on Planetary Defense*, pg. 11.
[22] Urias, *Planetary Defense,* pg. 8.
[23] Hamilton, Solar Systems Collisions Simulator.
[24] New International Year Book, 1910, pg. 640.
[25] Britain Nuclear Weapon Testing Website.
[26] Samuel Glasstone, Nuclear Weapons Effect Simulator.
[27] Urias, *Planetary Defense,* pg. 9.
[28] NASA *Spaceguard Survey,* pgs. 32-33.
[29] Morrison, Asteroid and Comet Impact Hazard News, NASA Website
[30] Chamberlin, NEO Statistics, NASA Website
[31] Chamberlin, NEO Statistics, NASA Website
[32] Morrison, Asteroid and Comet Impact Hazard News, NASA Website
[33] NASA *Spaceguard Survey*, pgs. 26-27.
[34] Chamberlin, NEO Statistics, NASA Website
[35] Schweickart, *NEOs: The Katrinas of the Cosmos?*
[36] NASA ARC NEO News Status Report Online
[37] Knox, *Planetary Defense Legacy for a Certain Future,* pg. 14.
[38] United States Government, FY 2008 budget.
[39] NEO Interception Workshop Summary Report, pg. 30-31.
[40] Samuel Glasstone, Nuclear Weapons Effect Simulator.
[41] Kowitz, *The Department of Defense Take on Planetary Defense*, pg. 10.
[42] NEO Interception Workshop Summary Report, pgs. 24-25.
[43] NEO Interception Workshop Summary Report. pgs. 24-25.
[44] NEO Interception Workshop Summary Report, pgs. 17-21.
[45] Urias, *Planetary Defense,* pg. 54.
[46] Galvez, ESA Website
[47] NASA Clementine Website
[48] Urias, *Planetary Defense,* pg. 11.
[49] Urias, *Planetary Defense,* pg. 11.
[50] NEO Interception Workshop Summary Report, pgs. 25-26, 39-46.
[51] NEO Interception Workshop Summary Report, pg. 31.
[52] NEO Interception Workshop Summary Report, pg. 31.
[53] Space Launch Sites Website

Bibliography

"Britains Nuclear Weapons: Britain's Nuclear Testing." UK Ministry of Defense, 1998.
http://nuclearweaponarchive.org/Uk/UKTesting.html

Chamberlin, Alan. National Aeronautics and Space Administration, Near Earth Object Program.
"Near Earth Object Discovery Statistics" http://neo.jpl.nasa.gov/stats/

"Clementine Project Information" NASA Goddard Space Flight Center Data. Greenbelt,
Maryland, http://nssdc.gsfc.nasa.gov/planetary/clementine.html

Colby, Frank Moore and Churchhill, Allen Leon, "New International Yearbook, a Compendium
of the World's Progress for the Year 1909." copyright 1910. books.google.com/books

Gálvez, Andres. "ESA selects targets for asteroid-deflecting mission Don Quijote" European
Space Agency Press Release 412005http://www.esa.int/esaCP/SEML9B8X9DE_index_0

Hamilton, Calvin, J. "Terrestrial Impact Craters" copyright 1997-2001
http://www.solarviews.com/eng/tercrate.htm

Hamilton, Douglas, P. Dr. "Solar System Collisions" Model, Astronomy Workshop.
http://janus.astro.umd.edu/astro/impact/, January 2000

Harvard-Smithsonian Center for Astrophysics. Cambridge, Massachussetts. "List Of The
Potentially Hazardous Asteroids (PHAs)"
http://www.cfa.harvard.edu/iau/lists/Dangerous.html

Knox, James S. "Planetary Defense Legacy for a Certain Future" Air War College, Air
University, Maxwell Air Force Base, Alabama. April 1998.

Kowitz, Charles, H. "The Department of Defense Take on Planetary Defense: US Space Policy
for Defending Against Near-Earth Objects" Air Command and Staff College, Air
University, Maxwell Air Force Base, Alabama. April 2002.

Lewis, John, S. "Rain of Iron and Ice: The Very Real Threat of Comet and Asteroid
Bombardment" Addison Wesley Publishing Company, New York, 1995.

Meteor Crater Enterprises/Science Data Corporation. "The First Proven and Best Preserved
Meteorite Crater on Earth." http://www.meteorcrater.com/index.php copyright 2006

Morrison, David "Asteroid and Comet Impact Hazards" Archive of News, NASA Ames
Research Center Website. http://impact.arc.nasa.gov/news_detail.cfm?ID=172

NASA ARC NEO News "Status Report March 7, 2007" Ames Research Center. Planetary
  Defense Conference. http://www.earthtoday.net/news/viewsr.html?pid=23569

"Near Earth Object Interception Workshop Summary Report" Sponsored by NASA, held at Los
  Alamos, New Mexico. Edited by Anita Sohus, Jet Propulsion Laboratory, August 31st,
  1992.

"New International Year Book: A Compendium of the World's Progress" Edited by Frank
  Moore Colby, Dodd, Meade and Company, New York, 1911.

Samuel Glasstone and Philip J. Dolan., "The Effects of Nuclear Weapons" 3rd Edition
  http://meyerweb.com/eric/tools/gmap/hydesim.html

Schweickart, Russell L. "NEOs: The Katrinas of the Cosmos?" International Space Development
  Conference, 2006. Luncheon Address on May 6, 2006
  www.b612foundation.org/papers/ISDC06.doc

Sharpton, Virgil L. "Meteor." World Book Online Reference Center. 2005. World Book, Inc.
  http://www.worldbookonline.com/wb/Article?id=ar358150

"Space and Rocket Launch Sites Around the World" Space Today Online, 2004
  http://www.spacetoday.org/Rockets/Spaceports/LaunchSites.html

Spaceguard Survey, Asteroid and Comet Impact Hazards, NASA Ames Space Science Division
  http://impact.arc.nasa.gov/downloads/spacesurvey.pdf

The Barringer Crater Company. "What is the Barringer Meteorite Crater?"
  http://www.barringercrater.com/

United States Government "Budget of the United States Governmnet, Fiscal Year 2008"
  http://www.whitehouse.gov/omb/budget/fy2008/

Urias, John, M. "Planetary Defense: Catastrophic Health Insurance for Planet Earth" Air Force
  2025 Project, Published October 1996.

Williams, David R. "Clementine Project Information" NASA Goddard Space Flight Center.
  http://nssdc.gsfc.nasa.gov/planetary/clementine.html

Young, Kelly. "Chances Small for Head on Collision with Killer Asteroid", Florida Today,
  September 1, 2002. http://www.space.com/scienceastronomy/fl_side2_020901.html

Illustrations

Figure 1

No Hazard (White Zone)	0	The likelihood of a collision is zero, or is so low as to be effectively zero. Also applies to small objects such as meteors and bodies that burn up in the atmosphere as well as infrequent meteorite falls that rarely cause damage.
Normal (Green Zone)	1	A routine discovery in which a pass near the Earth is predicted that poses no unusual level of danger. Current calculations show the chance of collision is extremely unlikely with no cause for public attention or public concern. New telescopic observations very likely will lead to re-assignment to Level 0.
Meriting Attention by Astronomers (Yellow Zone)	2	A discovery, which may become routine with expanded searches, of an object making a somewhat close but not highly unusual pass near the Earth. While meriting attention by astronomers, there is no cause for public attention or public concern as an actual collision is very unlikely. New telescopic observations very likely will lead to re-assignment to Level 0.
	3	A close encounter, meriting attention by astronomers. Current calculations give a 1% or greater chance of collision capable of localized destruction. Most likely, new telescopic observations will lead to re-assignment to Level 0. Attention by public and by public officials is merited if the encounter is less than a decade away.
	4	A close encounter, meriting attention by astronomers. Current calculations give a 1% or greater chance of collision capable of regional devastation. Most likely, new telescopic observations will lead to re-assignment to Level 0. Attention by public and by public officials is merited if the encounter is less than a decade away.

Threatening (Orange Zone)	5	A close encounter posing a serious, but still uncertain threat of regional devastation. Critical attention by astronomers is needed to determine conclusively whether or not a collision will occur. If the encounter is less than a decade away, governmental contingency planning may be warranted.
	6	A close encounter by a large object posing a serious but still uncertain threat of a global catastrophe. Critical attention by astronomers is needed to determine conclusively whether or not a collision will occur. If the encounter is less than three decades away, governmental contingency planning may be warranted.
	7	A very close encounter by a large object, which if occurring this century, poses an unprecedented but still uncertain threat of a global catastrophe. For such a threat in this century, international contingency planning is warranted, especially to determine urgently and conclusively whether or not a collision will occur.
Certain Collisions (Red Zone)	8	A collision is certain, capable of causing localized destruction for an impact over land or possibly a tsunami if close offshore. Such events occur on average between once per 50 years and once per several 1000 years.
	9	A collision is certain, capable of causing unprecedented regional devastation for a land impact or the threat of a major tsunami for an ocean impact. Such events occur on average between once per 10,000 years and once per 100,000 years.
	10	A collision is certain, capable of causing global climatic catastrophe that may threaten the future of civilization as we know it, whether impacting land or ocean. Such events occur on average once per 100,000 years, or less often.

Figure 1 (continued)

Figure 2

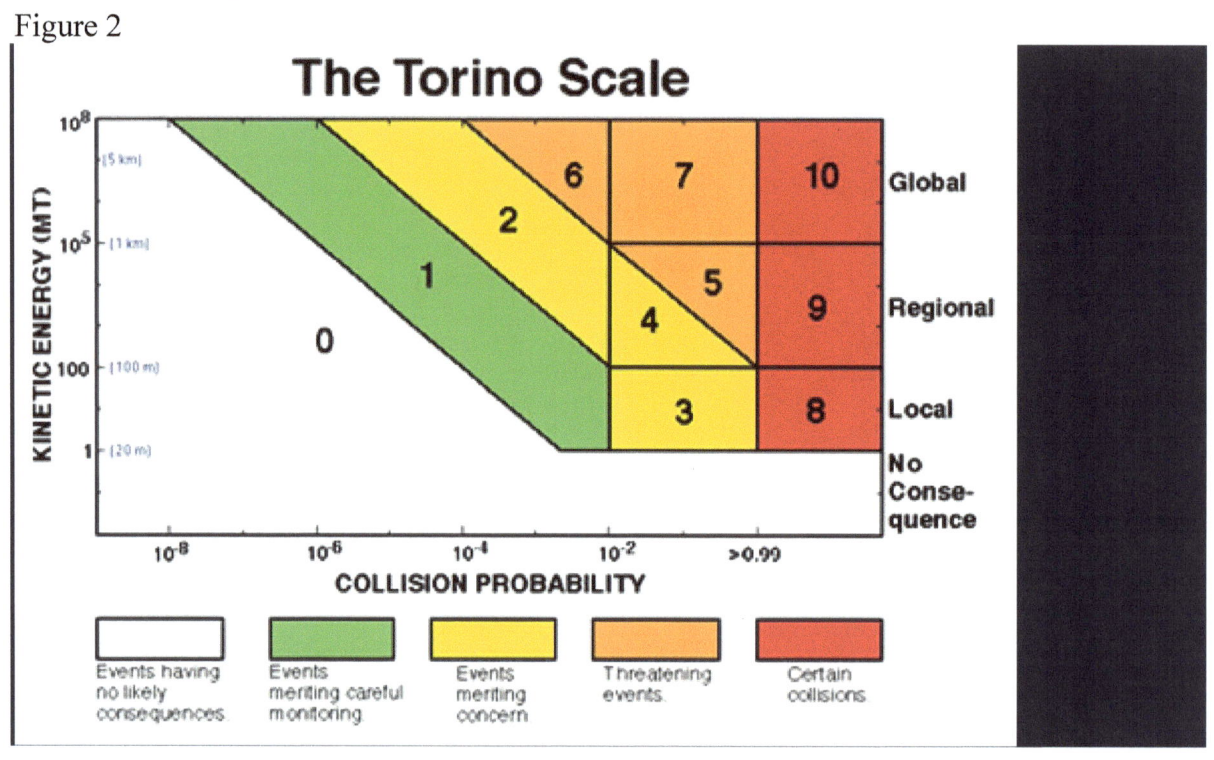

The Torino Scale

Figure 3

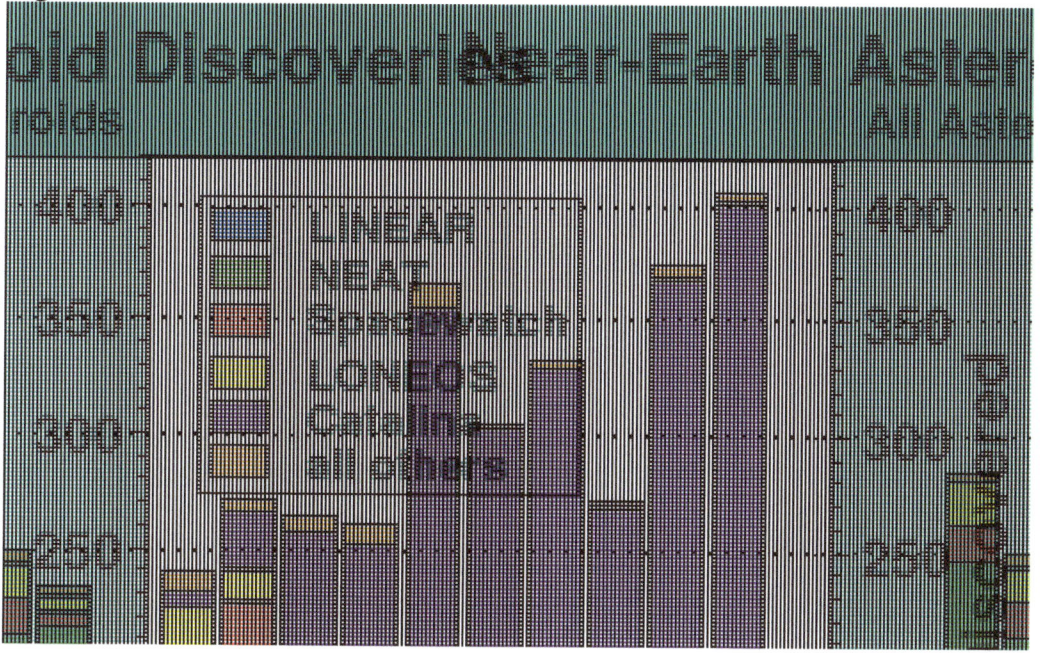

Figure 4

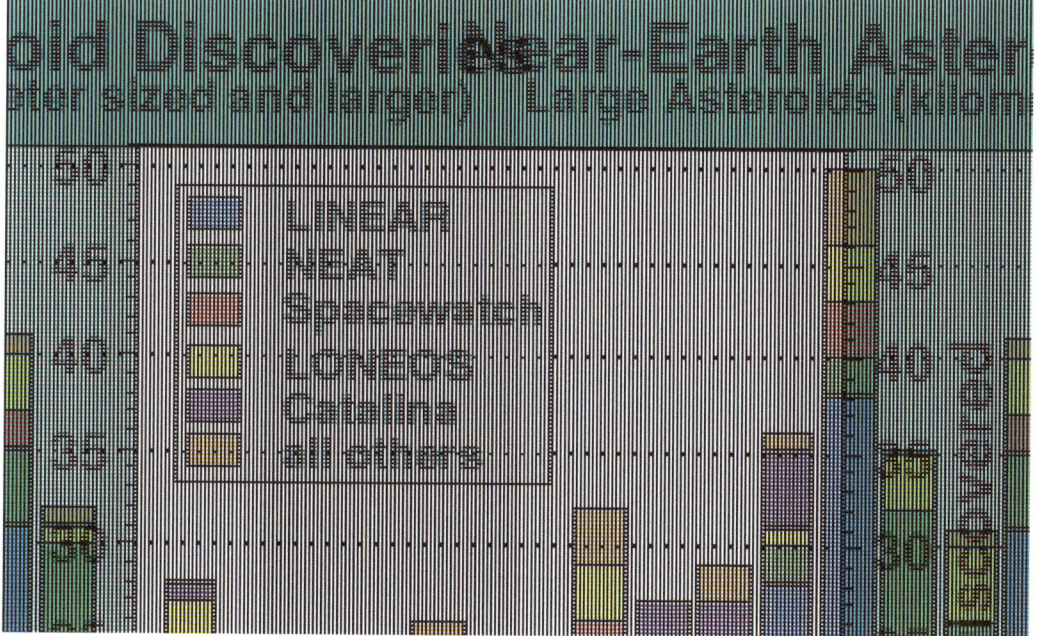

41

Figure 5

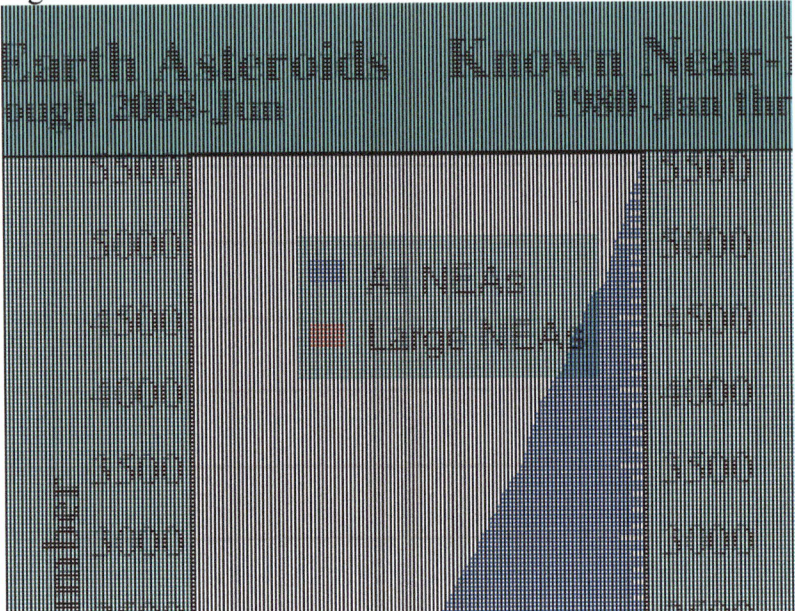

www.ingramcontent.com/pod-product-compliance
Lightning Source LLC
Chambersburg PA
CBHW050838180526
45159CB00004B/1953